DNA, God, & You!

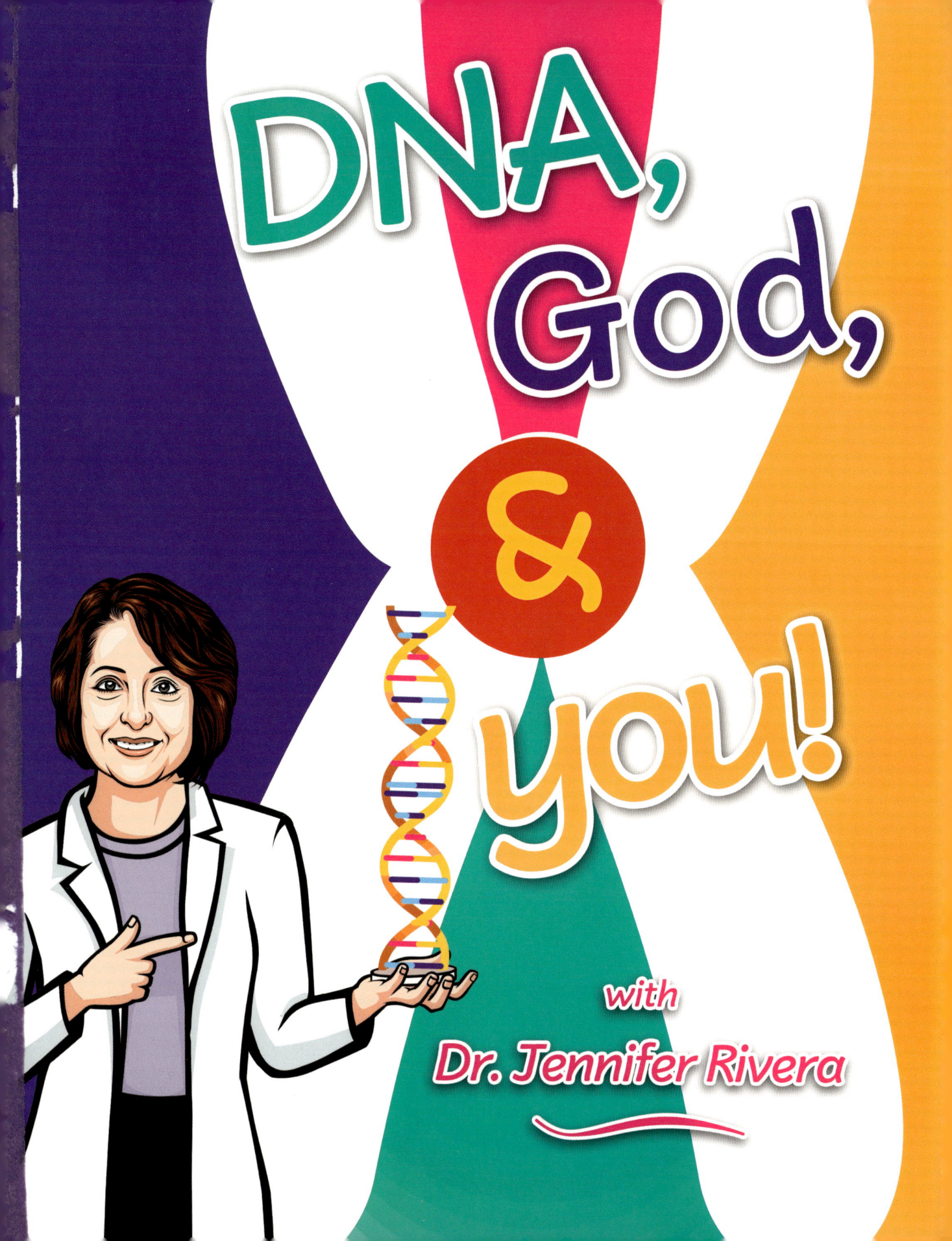

with Dr. Jennifer Rivera

First printing: November 2024

Copyright © 2024 by Jennifer Rivera and Master Books. All rights reserved. No part of this book may be used or reproduced in any manner whatsoever without written permission of the publisher, except in the case of brief quotations in articles and reviews. For information write:

Master Books, P.O. Box 726, Green Forest, AR 72638
Master Books® is a division of the New Leaf Publishing Group, LLC.

ISBN: 978-1-68344-375-9
ISBN: 978-1-61458-893-1 (digital)
Library of Congress Number: 2024947249

Scripture quotations are from the ESV® Bible (The Holy Bible, English Standard Version®), © 2001 by Crossway, a publishing ministry of Good News Publishers. Used by permission. All rights reserved. The ESV text may not be quoted in any publication made available to the public by a Creative Commons license. The ESV may not be translated in whole or in part into any other language.

Please consider requesting that a copy of this volume be purchased by your local library system.

Printed in the United States

Image credits:

All images are from Shutterstock.com unless otherwise noted.

Dr. Rivera illustrations by Sarat M.

Adobestock.com: p 17, p 21 8th image, p 27, p 28 right, p 29, p 30 top, p 31 top

NLPG staff: p 21 2nd

AiG: p 35, Pterosaur, T-rex, Wild Hog, ceratopsian, Adam and Eve.

Dedication

To my friend and colleague Dr. Georgia Purdom, a molecular geneticist, who believes the orderly, complex nature of DNA is one of the greatest confirmations of the biblical Creator God.

Discover:

- DNA and the Bible
- What does DNA stand for?
- When Life Begins
- The Numbers
- How DNA Makes You-nique!
- What about Twins?
- Replication & Genetic Code
- How DNA Is Designed
- T bonds with A, G goes with C
- DNA and our Creator God
- Animal Kinds – DNA?
- All Creation Has DNA

Hi, I am *Dr. Jennifer Rivera*, a forensic scientist. DNA is an important tool we use in criminal investigation. But what exactly is DNA? Join me as we take an inside look into the language of life called DNA, a complex code written by the Creator of the universe.

Fearfully and wonderfully made are you,
the Bible tells us this and we know it is true.
The Bible can be trusted from cover to cover,
for it is spoken by God; above Him is no other.

Ps 139 v 14

There are over 300 verses in the Bible that talk about creation and God as the Creator.

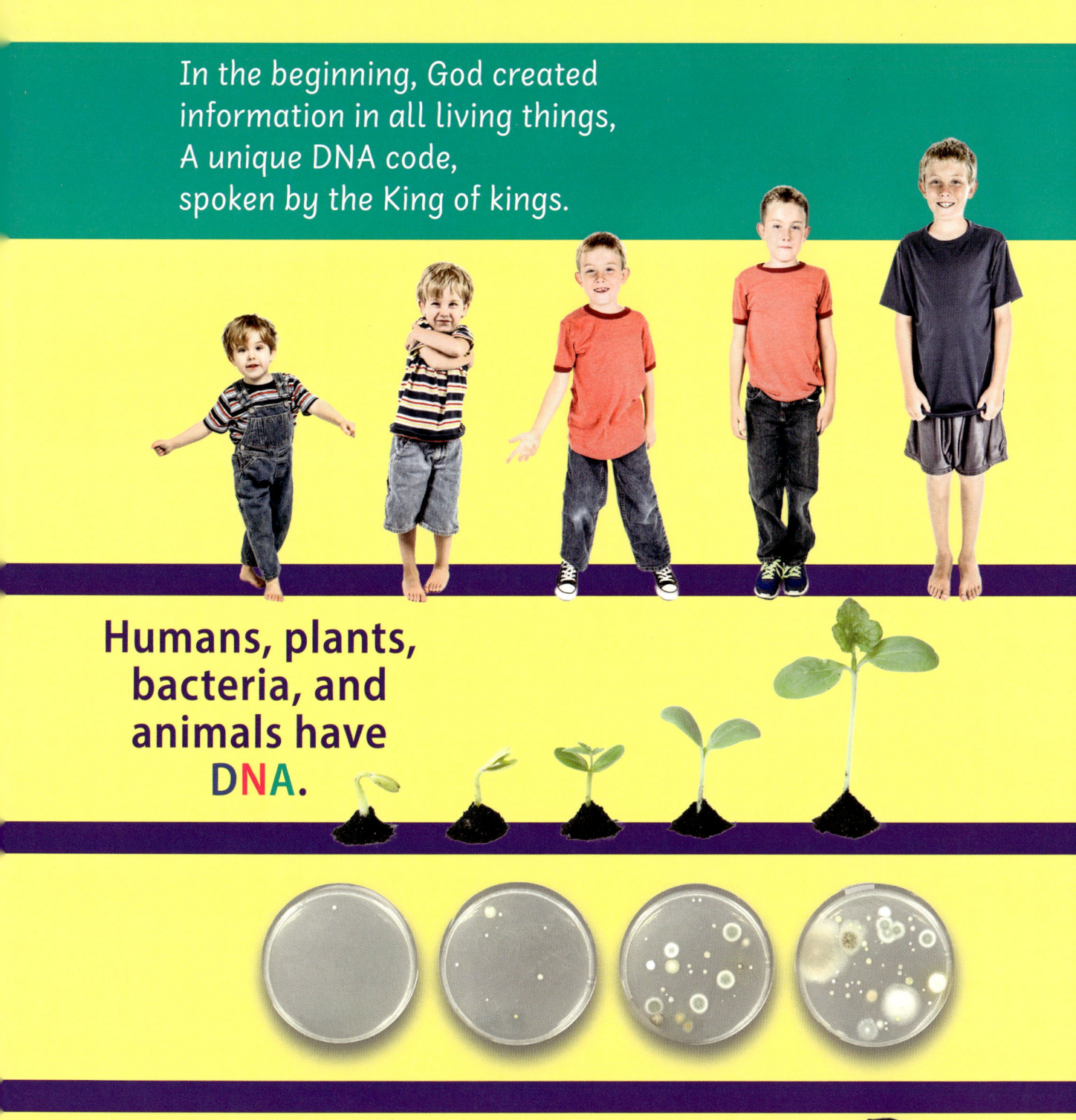

In the beginning, God created
information in all living things,
A unique DNA code,
spoken by the King of kings.

Humans, plants, bacteria, and animals have DNA.

"DNA is a molecule which contains the biological instructions that make each species unique."

—Human Genome Project

What days of CREATION include living things?

1
2
3

Light and earth

Separated waters above and below

Dry land and plants

6

4 5 6

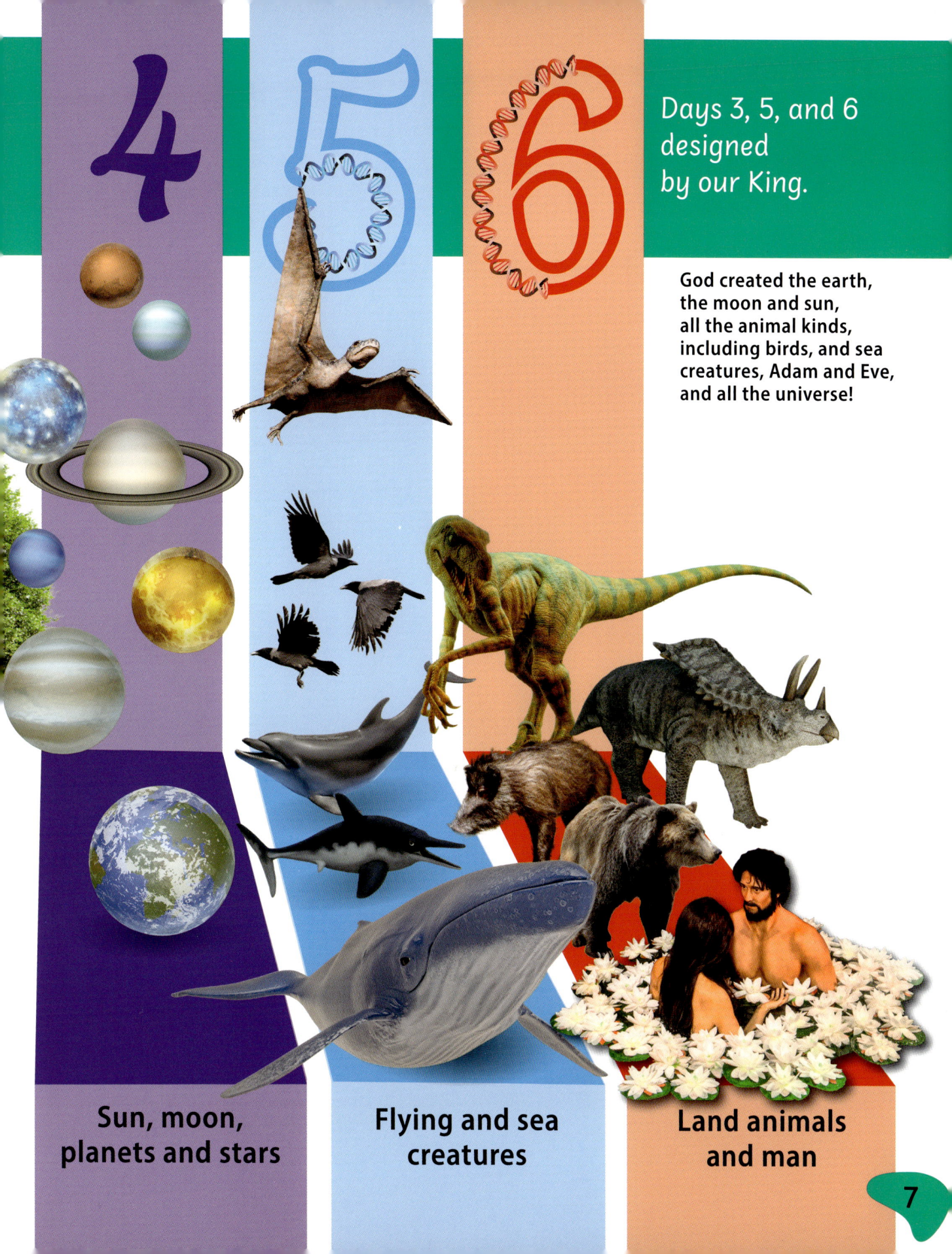

Days 3, 5, and 6 designed by our King.

God created the earth, the moon and sun, all the animal kinds, including birds, and sea creatures, Adam and Eve, and all the universe!

Sun, moon, planets and stars

Flying and sea creatures

Land animals and man

7

Scientists have been able to recover sequences of DNA from around 20 extinct animals like the woolly mammoth, woolly rhinoceros, passenger pigeon, cave bear, and Tasmanian tiger. Soft tissue-like cells and proteins have also been found in dinosaurs. DNA and soft tissue could not survive hundreds of thousands to millions of years of evolutionary time.

bak-TEER-ee-uh

Bacteria (bak-TEER-ee-uh) are tiny, single-celled organisms that get nutrients from their environments.

8

1 Timothy 6:13
tells us God gives life to all things.
All living things contain a specific DNA code —
bacteria, butterflies, animals, and even toads.

Fern cells under the microscope

1 Timothy 6:13
I charge you in the presence of God, who gives life to all things …

In other words...
God created life.

9

Dogs, cats, hamsters, and newts,

Every creature has DNA and every creature is unique.

Proverbs 12:10
Whoever is righteous has regard for the life of his beast, but the mercy of the wicked is cruel.

In other words...
Be kind to animals.

Plants, trees, flowers, and jutes.

Currently, fork ferns have the biggest genome at about 160 billion base pairs.

The smallest genome found in plants is the carnivorous corkscrew plant at 61 million base pairs.

11

DNA science confirms animals reproduce within their created kinds,
Just as Genesis chapter one says the word "kind" ten times.

[Original Created Cat Kind]

What is a Kind?

A kind is a group of animals or plants that can reproduce with another of the same kind.

Genetic Potential

[Lion]

[Jaguar]

[Cheetah]

[House Cat]

Because there were kinds, it meant that Noah's ark didn't have to take two of every animal. He only needed two representatives of each kind.

12

Genesis
1:11–12

And God said, "Let the earth sprout vegetation, plants yielding seed, and fruit trees bearing fruit in which is their seed, each according to its **kind**, on the earth." And it was so. The earth brought forth vegetation, plants yielding seed according to their own **kinds**, and trees bearing fruit in which is their seed, each according to its **kind**. And God saw that it was good. And there was evening and there was morning, the third day.

1:20–24

And God said, "Let the waters swarm with swarms of living creatures, and let birds fly above the earth across the expanse of the heavens." So God created the great sea creatures and every living creature that moves, with which the waters swarm, according to their **kinds**, and every winged bird according to its **kind**. And God saw that it was good. And God blessed them, saying, "Be fruitful and multiply and fill the waters in the seas, and let birds multiply on the earth." And there was evening and there was morning, the fifth day.

And God said, "Let the earth bring forth living creatures according to their **kinds**—livestock and creeping things and beasts of the earth according to their **kinds**." And it was so. And God made the beasts of the earth according to their **kinds** and the livestock according to their **kinds,** and everything that creeps on the ground according to its **kind**. And God saw that it was good.

1:26–30

Then God said, "Let us make man in our image, after our likeness. And let them have dominion over the fish of the sea and over the birds of the heavens and over the livestock and over all the earth and over every creeping thing that creeps on the earth." So God created man in his own image, in the image of God he created him; male and female he created them. And God blessed them. And God said to them, "Be fruitful and multiply and fill the earth and subdue it, and have dominion over the fish of the sea and over the birds of the heavens and over every living thing that moves on the earth." And God said, "Behold, I have given you every plant yielding seed that is on the face of all the earth, and every tree with seed in its fruit. You shall have them for food. And to every beast of the earth and to every bird of the heavens and to everything that creeps on the earth, everything that has the breath of life, I have given every green plant for food." And it was so. And God saw everything that he had made, and behold, it was very good. And there was evening and there was morning, the sixth day.

In other words...
God created everything. Why do all things on Earth, whether person, animal, or plant, have similar DNA? Because all were created by the same Creator God.

Where is DNA in our body?

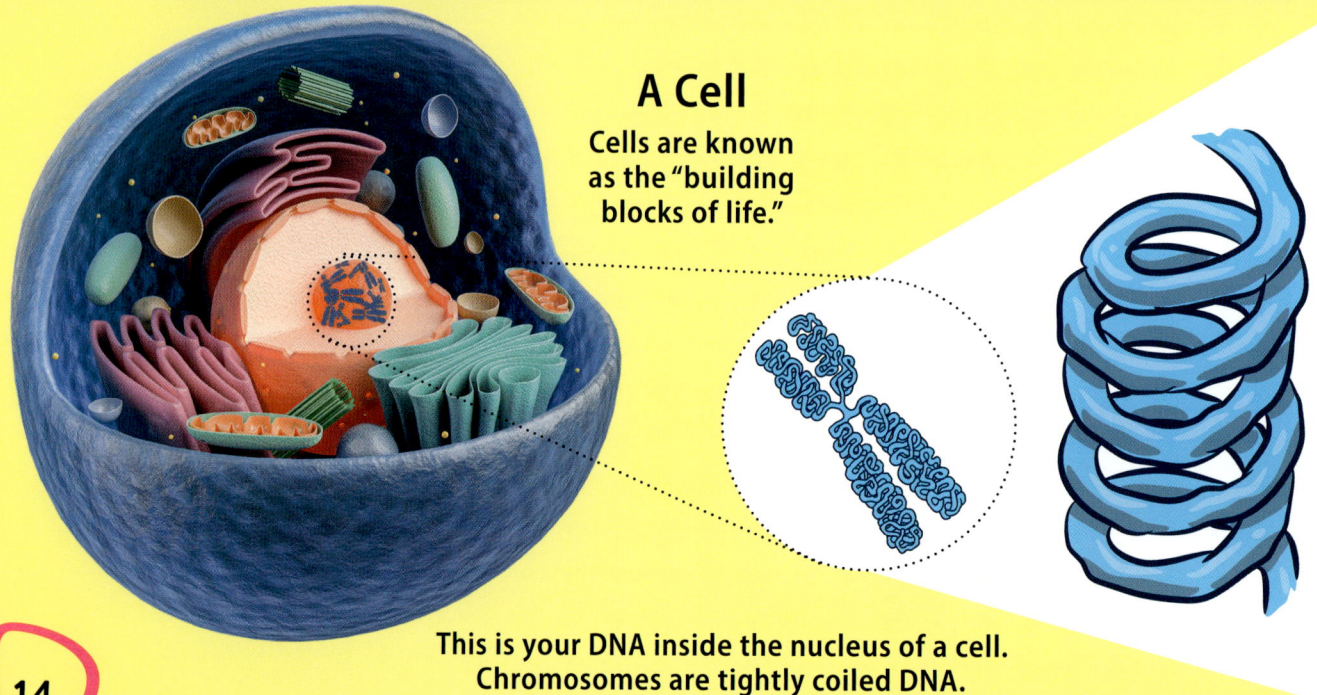

DNA is found in the nucleus of your cells, Like a SPRING, tightly wound into your chromosomes as well.

A Cell
Cells are known as the "building blocks of life."

This is your DNA inside the nucleus of a cell. Chromosomes are tightly coiled DNA.

14

Chromosomes are like ingredients in a cake.

Have you ever wondered how to make a delicious cake? To make a cake, we pair wet and dry ingredients.

We mix them all together in just the right way. Then, we pop the cake into the oven to bake at the perfect temperature for just the right amount of time.

Did you know that God puts us together in a special way too? Just like a cake has its ingredients, we have our own ingredients that make us unique. These ingredients are called CHROMOSOMES, and they make each of us one-of-a-kind. It's amazing how we're all put together just right, like a special recipe!

Wet ingredients

milk · eggs · eggs · vanilla · butter

Dry ingredients

salt · flour · sugar · baking powder

15

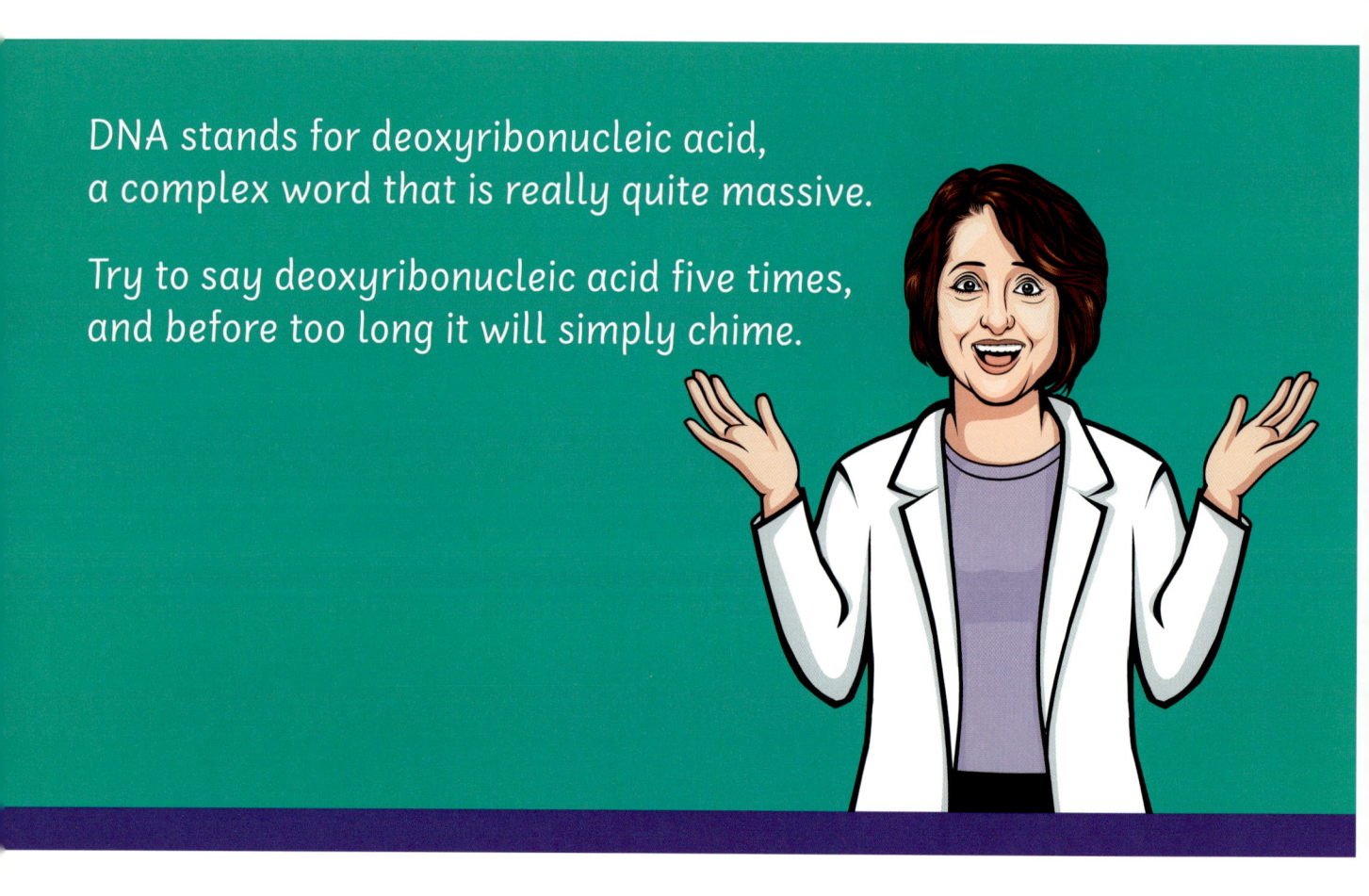

DNA stands for deoxyribonucleic acid,
a complex word that is really quite massive.

Try to say deoxyribonucleic acid five times,
and before too long it will simply chime.

What does DNA stand for?

DNA
<u>D</u>eoxyribo<u>N</u>ucleic <u>A</u>cid

16

Sound It Out!

Dee-ox-ee-rye-boh-noo-clay-ik As-id

DNA or Deoxyribonucleic Acid: A special recipe book put inside every living thing by God. It has all the instructions that tell our bodies how to grow, how we look, and more.

The base pairs are needed to help form the double helix, and its structure helps to replicate genetic information accurately.

The twisted shape of DNA is called a **double helix**.

Pairs of **bases** run along the inside of DNA.

How does DNA confirm the Creator God?

- DNA is a language.
- DNA is complex.
- DNA carries organized information.

The Bible tells us in Psalms
God knit us together;
Woven in the womb,
we are image bearers of the Creator.

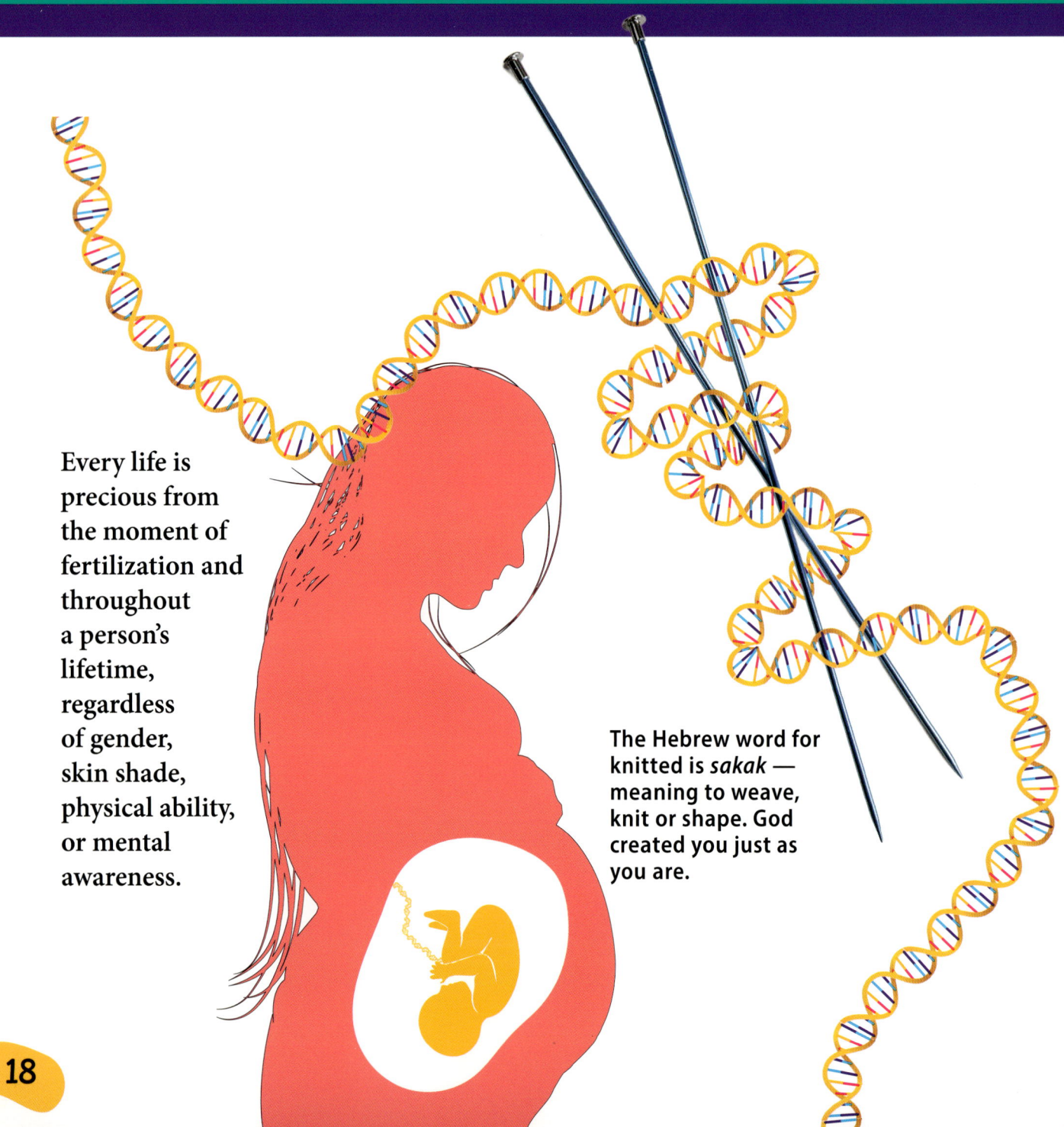

Every life is precious from the moment of fertilization and throughout a person's lifetime, regardless of gender, skin shade, physical ability, or mental awareness.

The Hebrew word for knitted is *sakak* — meaning to weave, knit or shape. God created you just as you are.

Your genetic information is complete at the moment of fertilization, One single cell, developing throughout life into the trillions.

DAY 2 — 2-CELL STAGE
DAY 3 — 4-CELL STAGE / 8-CELL STAGE
DAY 4
DAY 5
DAY 6–7
DAY 1
ZYGOTE

When life begins.

- Human life begins on DAY 1, there is 6 ft. (2m.) of DNA.
- Human life starts as a single cell known as a zygote.
- This single cell contains all the genetic information necessary for the development of a new individual. No new genetic information is added after fertilization.

Sound It Out!
zye-goat

Zygotes are formed through fertilization. The dad and the mom both share their DNA. The DNA has all the information for a baby to be formed.

A zygote is .05 inches. Can you see the white dot on this penny? ¾ inch

God created you with a DNA pattern divinely set,
A special sequence only you get.

Your core DNA sequence will remain the same throughout your lifetime if you remain healthy. Sometimes small changes may occur that affect how genes are expressed, but they will not affect the DNA sequence. These changes can occur from certain foods, medications, or toxins. There is also the possibility that mutations, or mistakes, in the DNA code can result in changes in the DNA sequence. Mutations are sometimes unnoticeable and other times quite harmful. Mutations are never the result of a DNA sequence acquiring new information, but the result of a change in preexisting information.

DNA will be important throughout your entire lifetime!

> **Psalm 139:13–14**
> For you formed my inward parts; you knitted me together in my mother's womb. I praise you, for I am fearfully and wonderfully made. Wonderful are your works; my soul knows it very well.

In other words...
God made you to be special and have a purpose. He has a plan for your life. Each one of us is unique and the workmanship of our Creator.

23 from mom and
23 from dad,
Added together is
46 chromosomes you have.

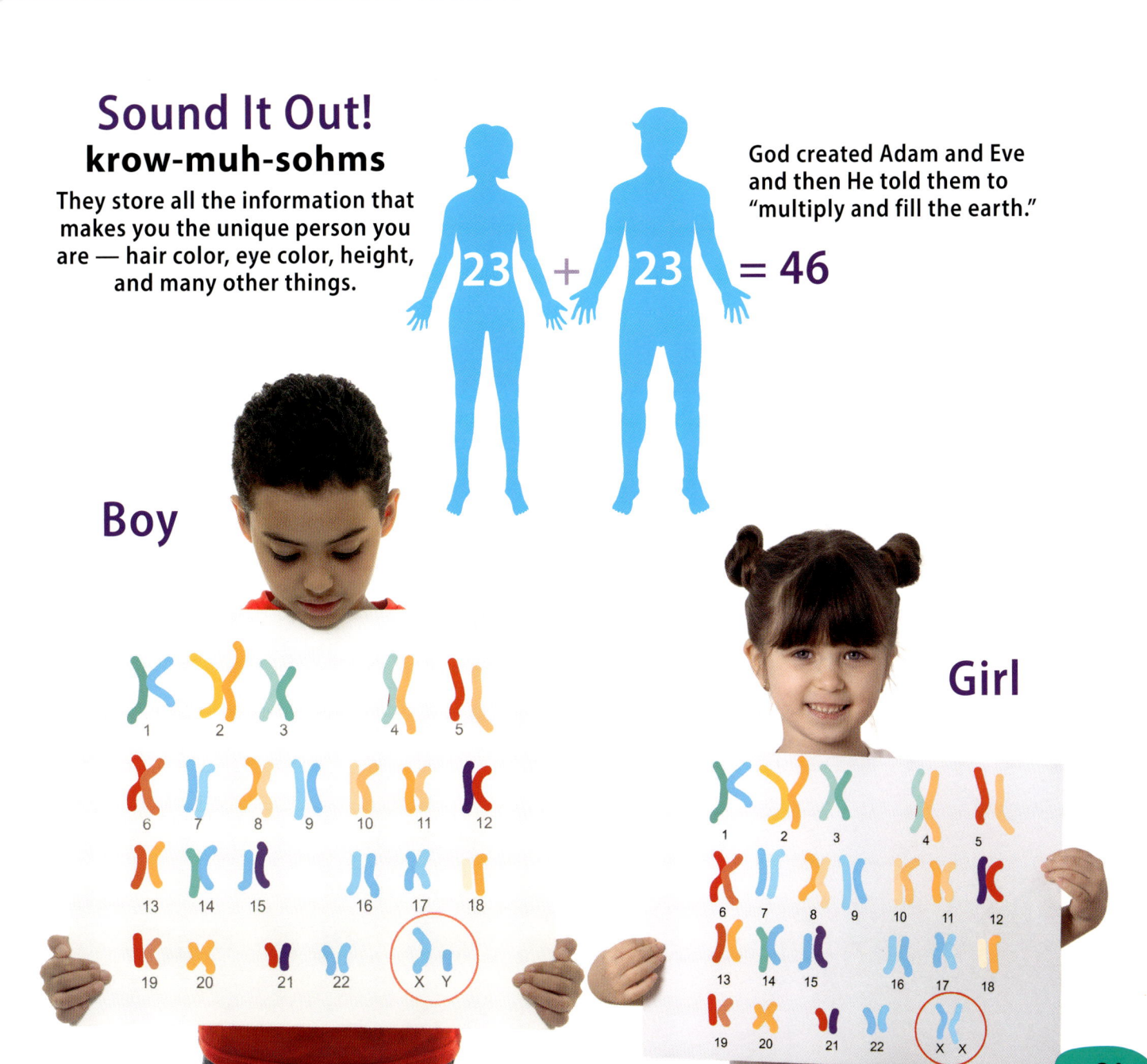

Sound It Out!
krow-muh-sohms

They store all the information that makes you the unique person you are — hair color, eye color, height, and many other things.

23 + 23 = 46

God created Adam and Eve and then He told them to "multiply and fill the earth."

Boy

Girl

21

Humans share
99% of their DNA code,
One human race,
like the Bible shows.

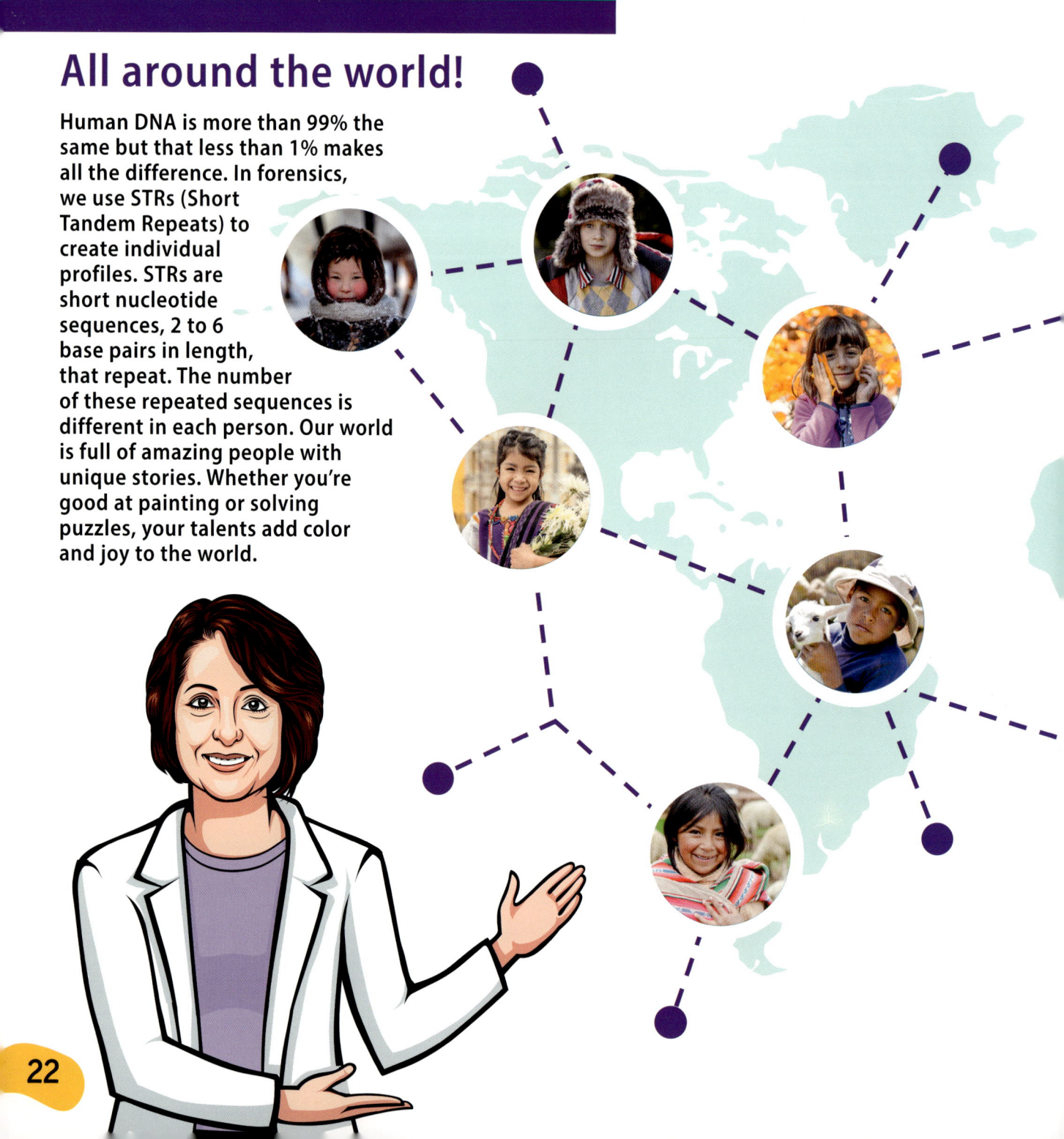

All around the world!

Human DNA is more than 99% the same but that less than 1% makes all the difference. In forensics, we use STRs (Short Tandem Repeats) to create individual profiles. STRs are short nucleotide sequences, 2 to 6 base pairs in length, that repeat. The number of these repeated sequences is different in each person. Our world is full of amazing people with unique stories. Whether you're good at painting or solving puzzles, your talents add color and joy to the world.

22

Acts 17:26
And he made from one man every nation of mankind to live on all the face of the earth, having determined allotted periods and the boundaries of their dwelling place …

In other words…
God made Adam and Eve, and from them came all the people, all over the earth! That is why our DNA is 99% the same.

DNA even holds clues to where some of your family may have lived a very, very long time ago.

23

The 23rd chromosome pair distinctly reveals whether you are a male or female. XY for boys and XX for girls is how you can tell.

From the moment of fertilization your gender, whether male or female, is revealed in your DNA.

Genesis 5:2
Male and female he created them, and he blessed them and named them Man when they were created.

In other words...
God created people as men and women, and He blessed them. He called them "mankind" when He made them.

Six feet of DNA is found in every nucleated cell, Among trillions of others in your body as well.

The average 10-year-old child has about 17 trillion cells. That's 17 followed by 12 zeros.

$$\begin{array}{r} 17{,}000{,}000{,}000{,}000 \text{ nucleated cells} \\ \times \quad 6 \text{ ft. of DNA per nucleated cell} \\ \hline = 24{,}000{,}000{,}000{,}000 \text{ feet of DNA in an} \\ \text{average 10-year-old child!} \end{array}$$

The average adult human male has around 36 trillion cells while adult females have 28 trillion.

As you grow, so does the number of cells you have.

If you lay 43,478 DNA helixes side by side, they would equal the width of an average human hair 100 micrometers wide.

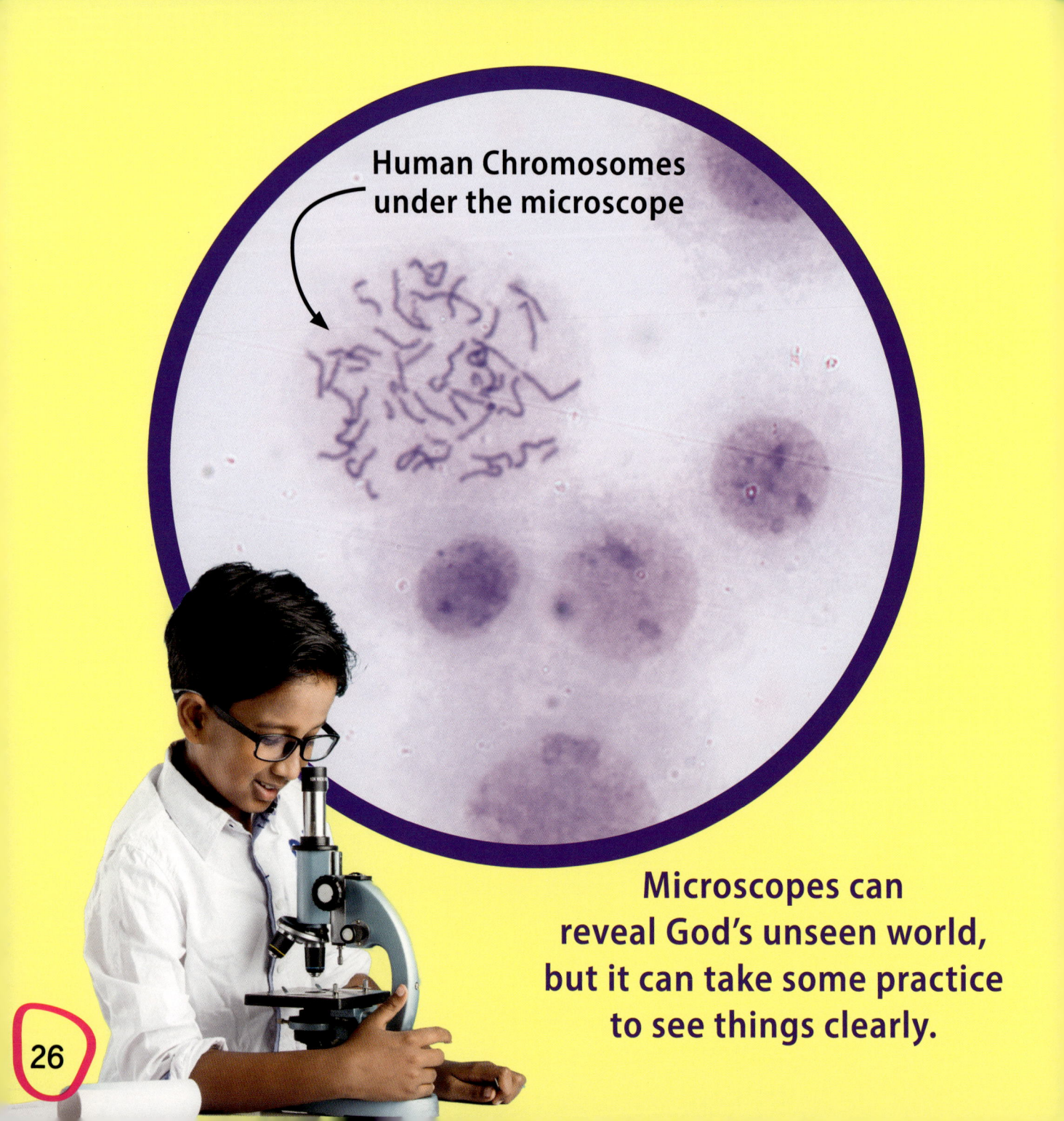

Human Chromosomes under the microscope

Microscopes can reveal God's unseen world, but it can take some practice to see things clearly.

Scientists use special microscopes to study DNA strands,
To help them understand the complexity of God's organized plans.

Cheek swab and blood both contain DNA.

DNA can be found in all biological fluids, such as saliva, sweat, white blood cells, and urine, as well as skin cells, hair follicles, and poop. Scientists have also found human DNA in mosquitoes.

27

DNA is made up of billions of instructions, information decoded through a process called transcription.

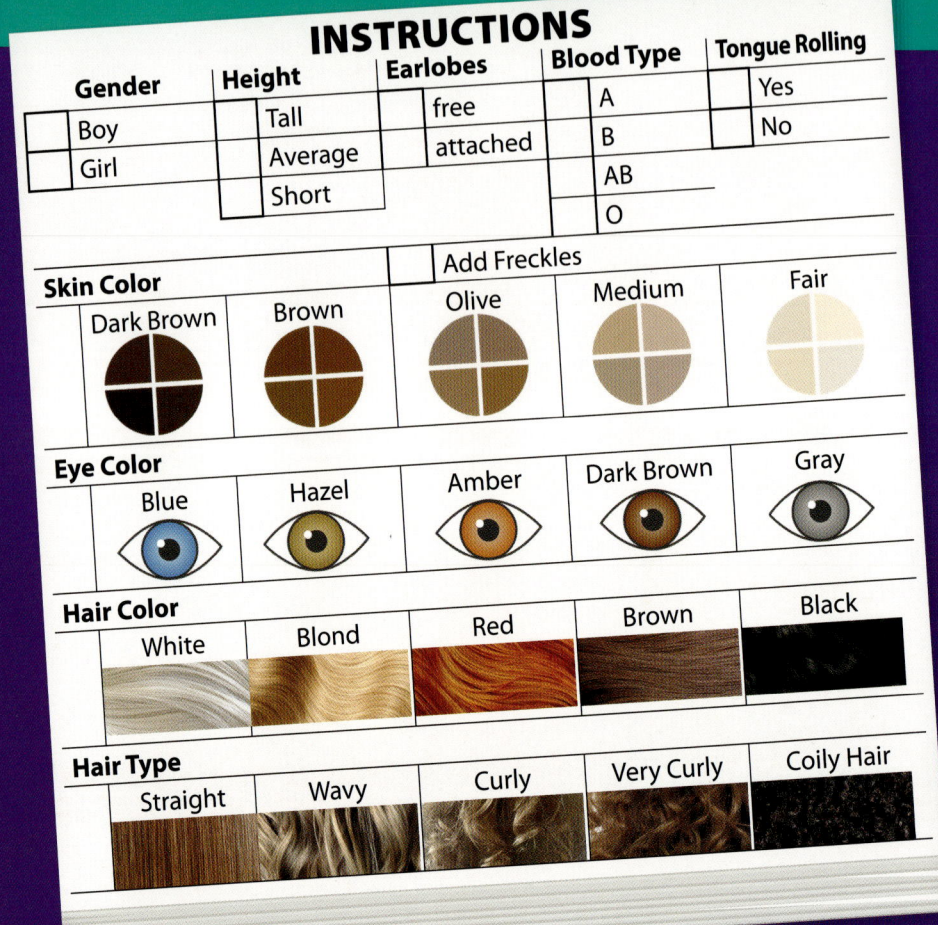

INSTRUCTIONS

Gender	Height	Earlobes	Blood Type	Tongue Rolling
Boy	Tall	free	A	Yes
Girl	Average	attached	B	No
	Short		AB	
			O	

Add Freckles

Skin Color: Dark Brown, Brown, Olive, Medium, Fair

Eye Color: Blue, Hazel, Amber, Dark Brown, Gray

Hair Color: White, Blond, Red, Brown, Black

Hair Type: Straight, Wavy, Curly, Very Curly, Coily Hair

DNA is made up of sections called genes.

Transcription is the copying of DNA sequences into messenger RNA (mRNA).

The mRNA carries the genetic instructions for making proteins.

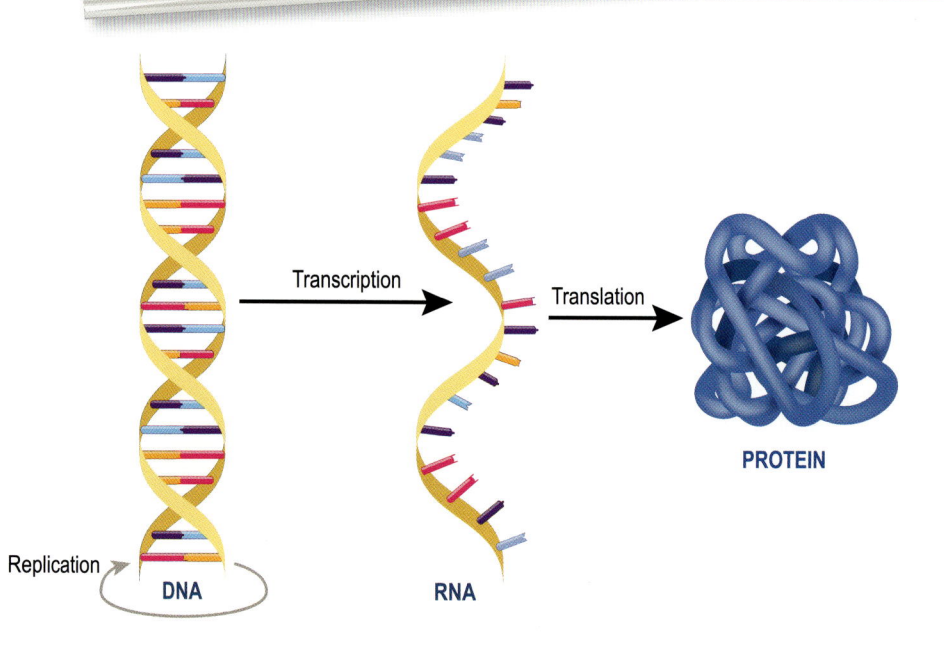

Replication — DNA → Transcription → RNA → Translation → PROTEIN

These instructions in genes contain characteristics and traits —
Your skin shade, eye color, and hair type, whether curly or straight.

Your freckles and ear shape, dimples and more,
Your eyebrows and chin,
and if you are too tall to walk through a door.

The gene expressing freckles is found on chromosome 4. Freckles are considered a dominant characteristic.

Helix

Antihelix

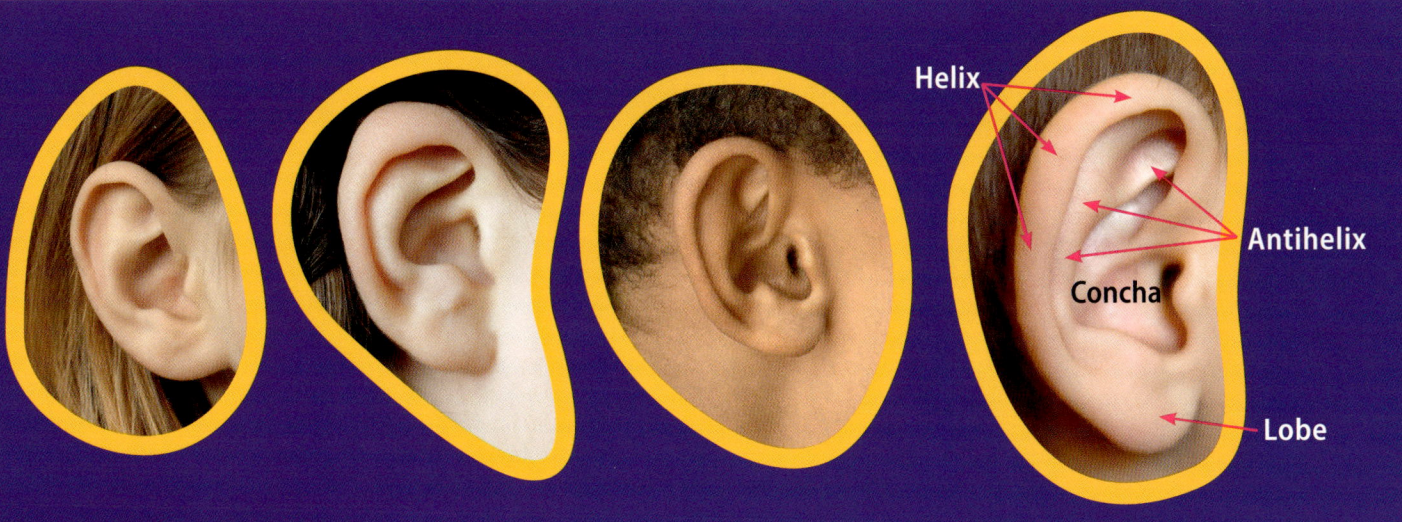

Concha

Lobe

Ear shape is a "fingerprint" just like DNA and fingerprints on your hands and feet. A technique was developed in 2011 called Cameriere's ear identification method. This method relies on measurements and ratios of different parts of the outside ear such as helix, antihelix, concha, and lobe. The likelihood of two individuals having the same ear shape code is less than 0.0007% when using the Cameriere's ear identification method.

Do ears stop growing?
YES THEY DO!

31

With over 3 billion base pairs in our 46 chromosomes,

TWIN

It's what makes people look different from their head to their toes.

Identical twins are the only ones who share 99.9% of their DNA information and look almost exactly the same, but they are each a unique person created by God with different personalities, interests, and talents.

DNA has many important functions, like coding information and DNA replication.

DNA replication

Old DNA strand

New DNA strands

Yellow is original DNA, red is the newly copied DNA information.

DNA replication is a complex, orderly process that exactly copies your existing DNA information. Complexity and orderly processes are only possible because an orderly God created the universe and spoke into existence the scientific laws we see throughout His creation.

The process of copying a DNA molecule to produce two identical DNA molecules is called DNA replication.

It has to happen before a cell can divide.

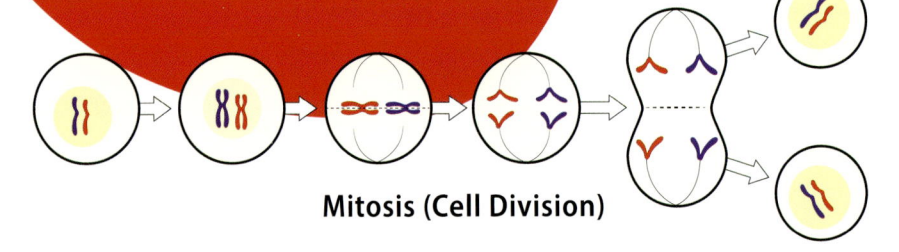

Mitosis (Cell Division)

34

DNA has organization and structure,
a complex language written by an Author who is our Creator.

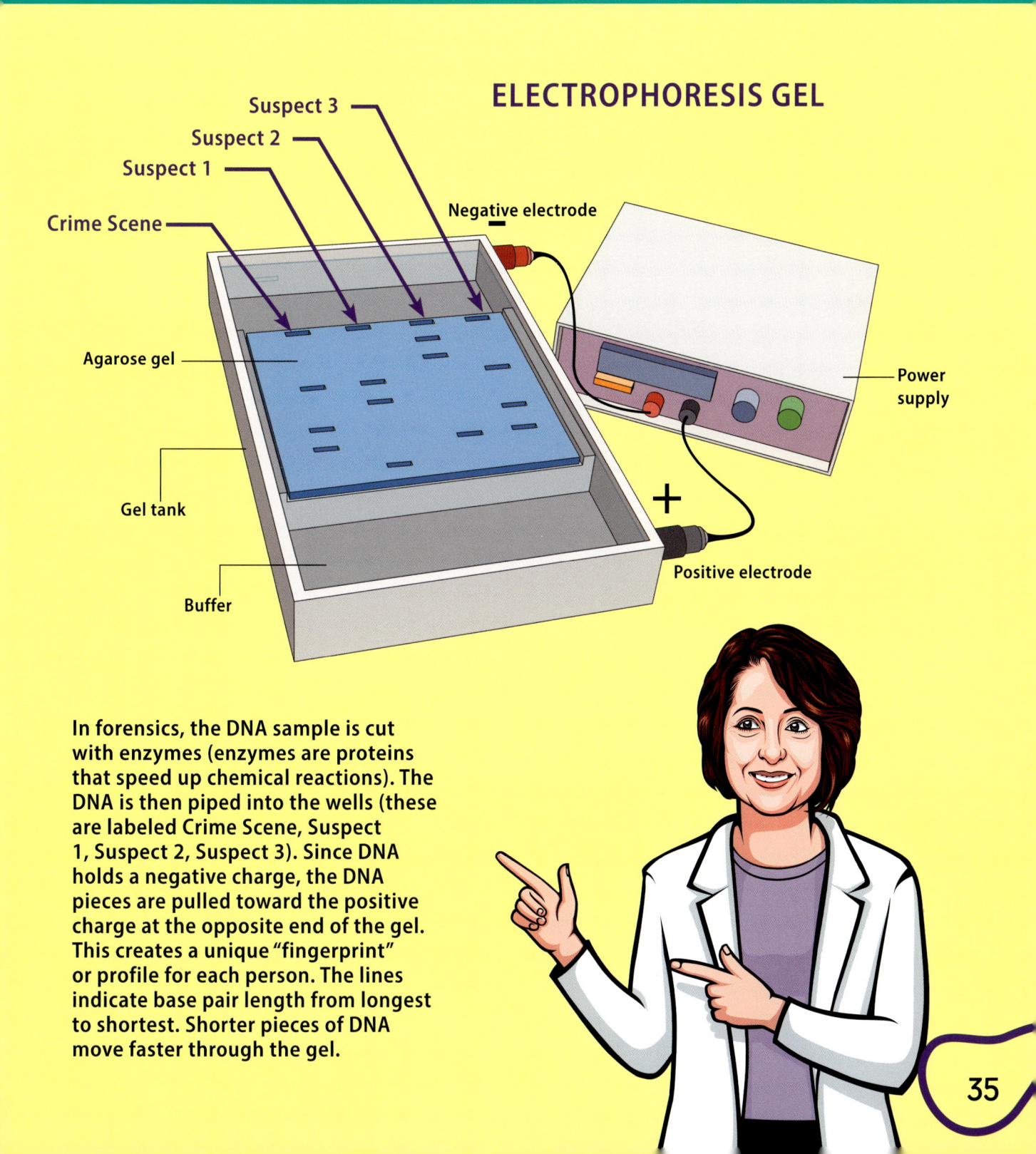

ELECTROPHORESIS GEL

- Suspect 3
- Suspect 2
- Suspect 1
- Crime Scene
- Negative electrode
- Agarose gel
- Gel tank
- Buffer
- Power supply
- Positive electrode

In forensics, the DNA sample is cut with enzymes (enzymes are proteins that speed up chemical reactions). The DNA is then piped into the wells (these are labeled Crime Scene, Suspect 1, Suspect 2, Suspect 3). Since DNA holds a negative charge, the DNA pieces are pulled toward the positive charge at the opposite end of the gel. This creates a unique "fingerprint" or profile for each person. The lines indicate base pair length from longest to shortest. Shorter pieces of DNA move faster through the gel.

DNA looks like a twisted ladder you can climb,
And you will learn its pattern in simply no time.

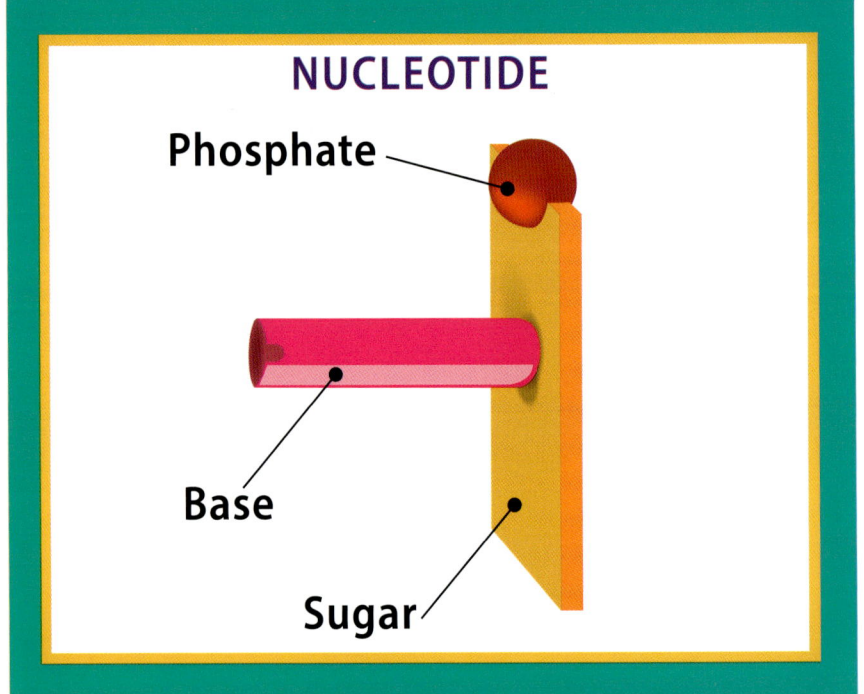

NUCLEOTIDE

- Phosphate
- Base
- Sugar

Made of 3 parts — a phosphate, sugar, and a nitrogen base — This complex design was created by God's grace.

Luke 18:27
But he said, "What is impossible with man is possible with God."

In other words...
God can make or do anything.

Four nitrogen bases make up DNA strands,
pairing together into God's divine plan.
Adenine (A) and Thymine(T) bond together like glue;
Guanine (G) and Cytosine (C) form a pair too.

Genetic Codes are instructions written into your DNA by the Creator God. The four bases, A, T, G, and C, combine to code for amino acids. Amino acids are the building blocks for proteins. There are thousands of proteins in the human body.

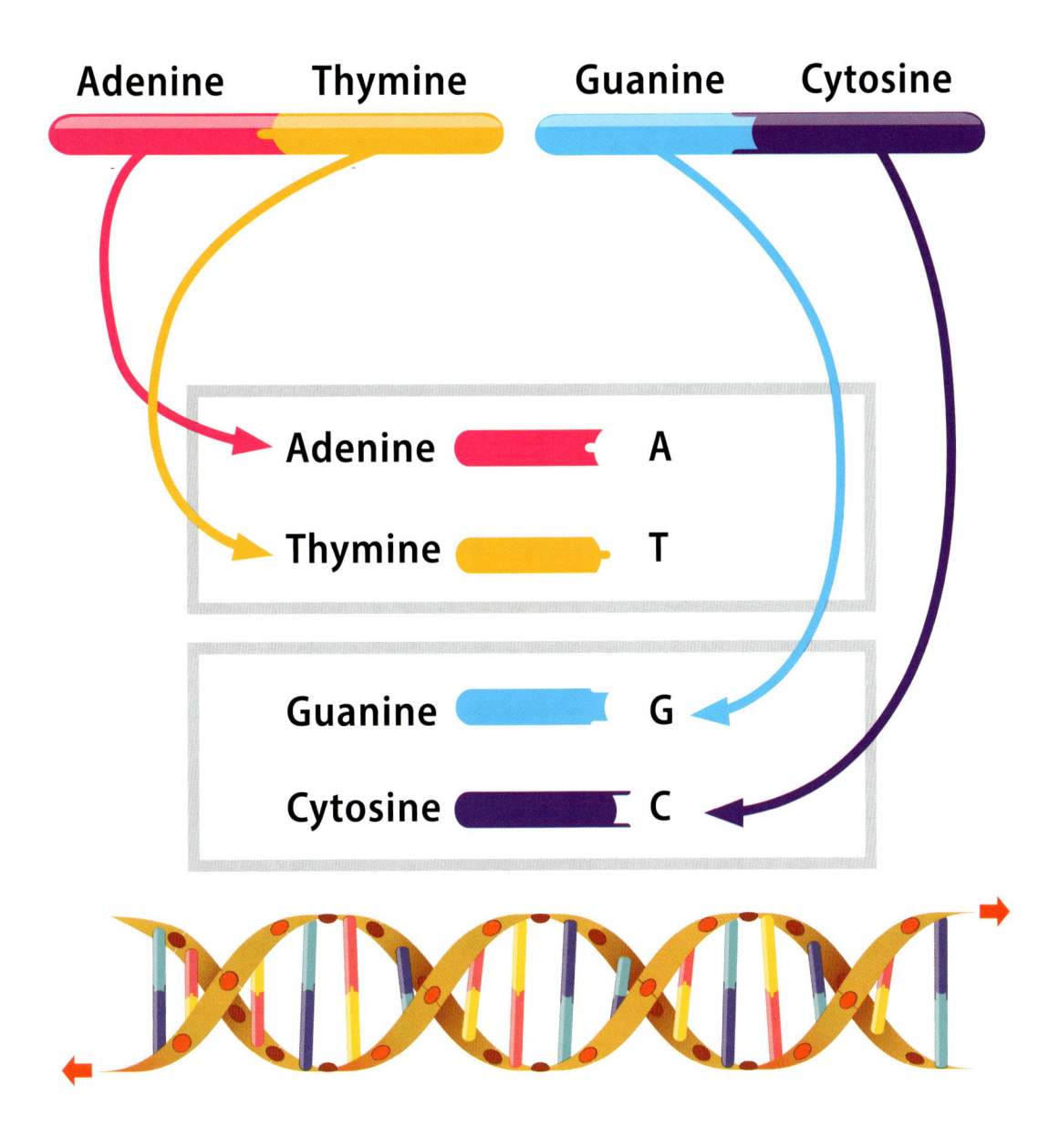

A hydrogen bond connects the base pairs, Creating rungs on the ladder, it is quite an affair.

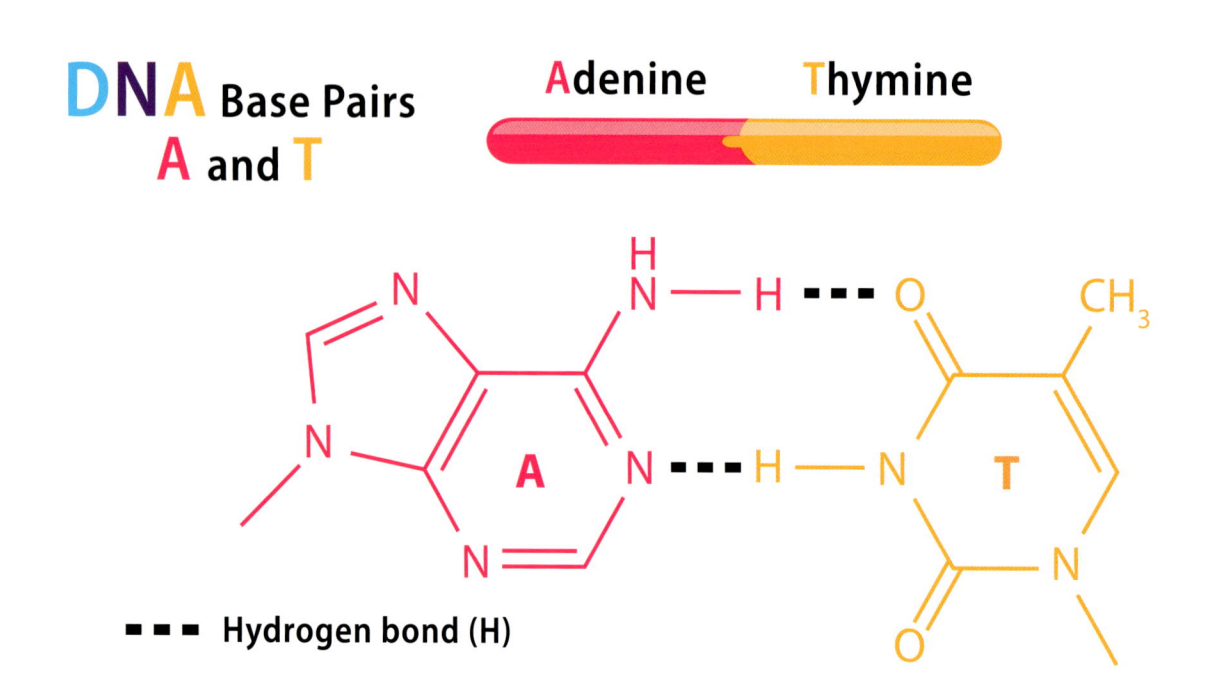

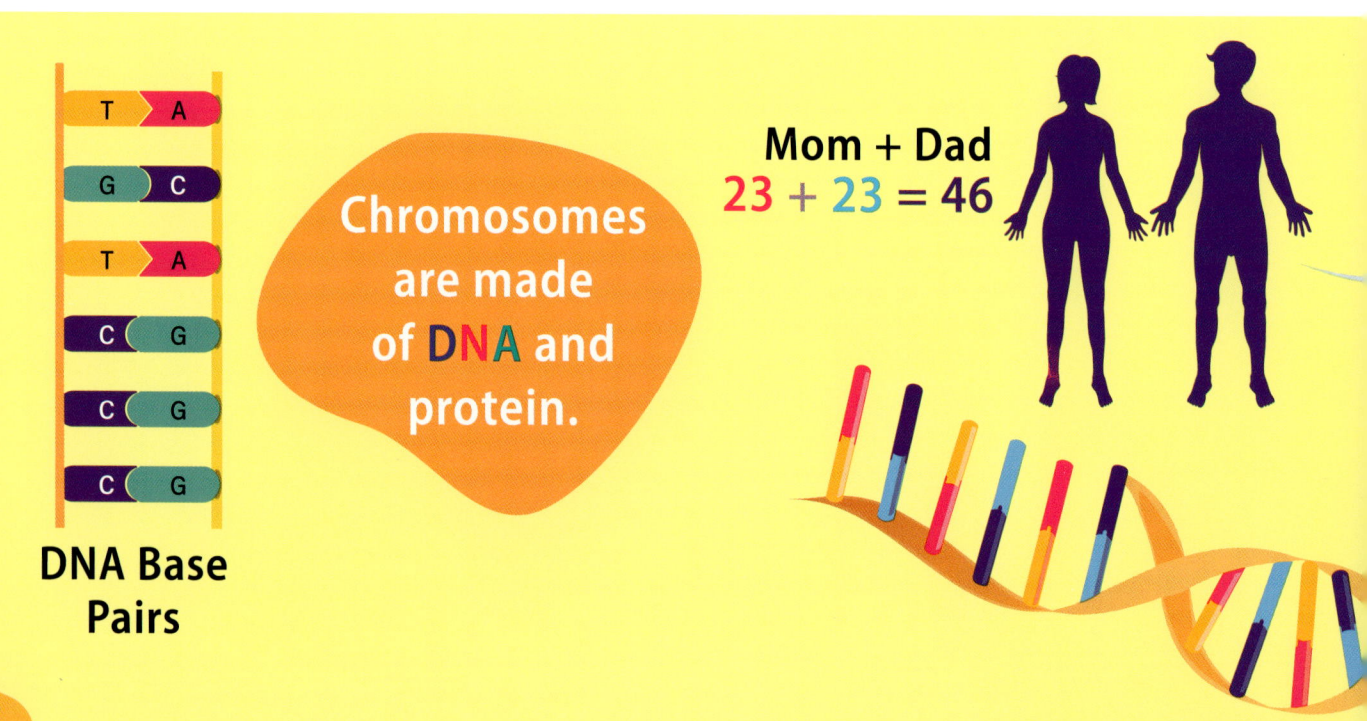

DNA Base Pairs A and T

Adenine — **Thymine**

- - - Hydrogen bond (H)

Chromosomes are made of DNA and protein.

DNA Base Pairs

Mom + Dad
23 + 23 = 46

38

T pairs with A on the DNA strand,
both with straight lines on top, making it grand.
G only pairs with C, as you will see,
both letters curved on top, it simply agrees.

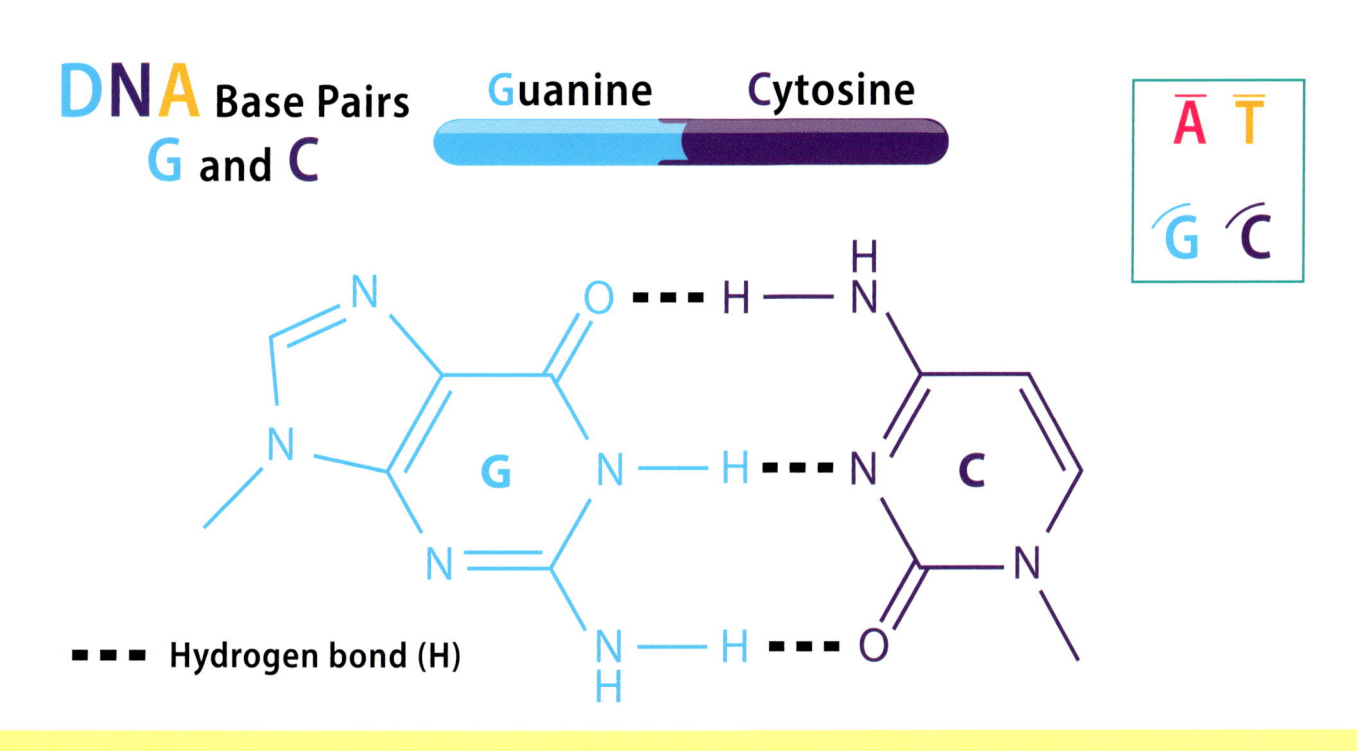

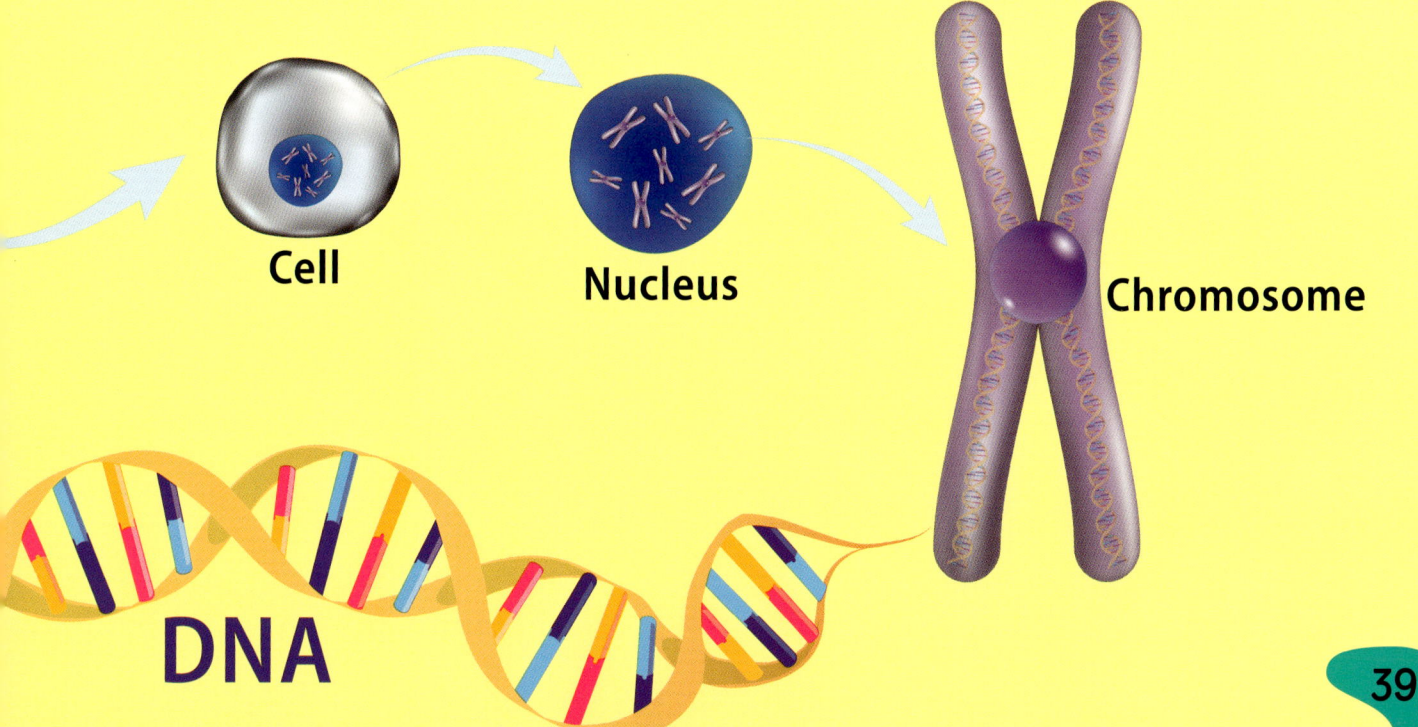

DNA Base Pairs G and C

Guanine   Cytosine

A T
G C

--- Hydrogen bond (H)

Cell → Nucleus → Chromosome

DNA

39

Within our genetic code
we see a lot of the same information,
Not just in animals and plants,
but people in every nation.

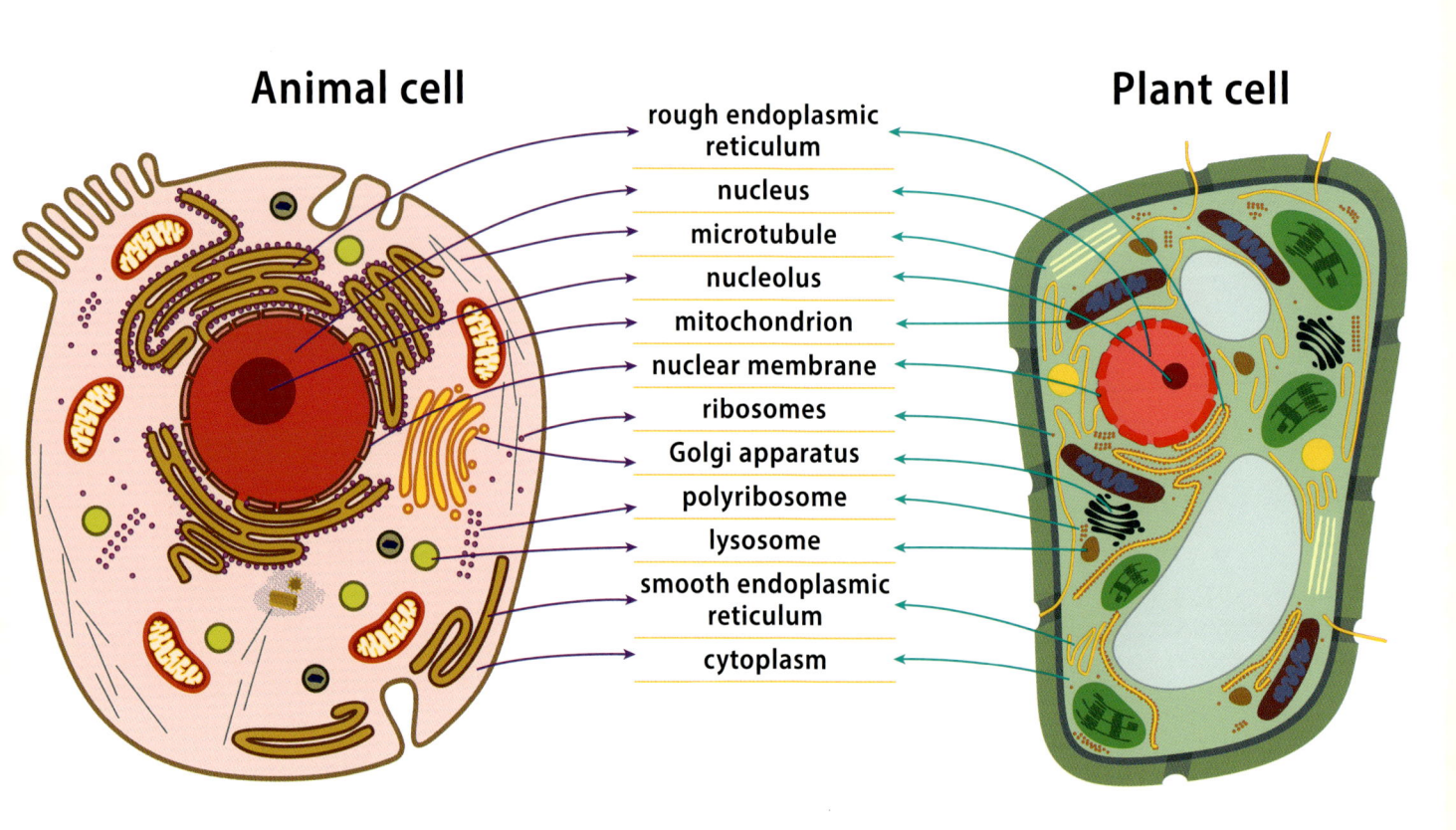

Animal cell

Plant cell

- rough endoplasmic reticulum
- nucleus
- microtubule
- nucleolus
- mitochondrion
- nuclear membrane
- ribosomes
- Golgi apparatus
- polyribosome
- lysosome
- smooth endoplasmic reticulum
- cytoplasm

God also created differences in plant and animal cells.

Plant cells have chloroplasts that allow the plant to photosynthesize and convert the sun's energy into chemical energy for the plant. If humans and animals had chloroplasts, we would be green from the chlorophyll!

Plant cells also have a cell wall that provides support and protection to the organelles inside the cell.

Remember! DNA is located in the nucleus of the cell!

40

What does this mean? Are we related to beans?

Many scientists consider plants to be "alive," but according to the Bible, plant life is not the same as animal and human life. The Hebrew word *nephesh* is used to distinguish human and animal life. When the Bible refers to mankind, *nephesh* means "living soul" and when it refers to animals, it means "living creature." These words are never used to describe plant life. Plants were created on day 3 of the creation week according to their kind. God instructed Adam, Eve, and the animals to eat plants for food.

https://answersingenesis.org/kids/science/are-plants-alive/

Genesis 1:29-30

And God said, "Behold, I have given you every plant yielding seed that is on the face of all the earth, and every tree with seed in its fruit. You shall have them for food. And to every beast of the earth and to every bird of the heavens and to everything that creeps on the earth, everything that has the breath of life, I have given every green plant for food." And it was so.

In other words...
God provides plants for food — for people and animals.

41

No! You are created by God,
uniquely human in design,
Given dominion over His earth,
in the fall, winter, spring, and summertime.

42

So, why do we see the same DNA throughout all living things?
A testimony to the same Creator God, to which all creation sings.

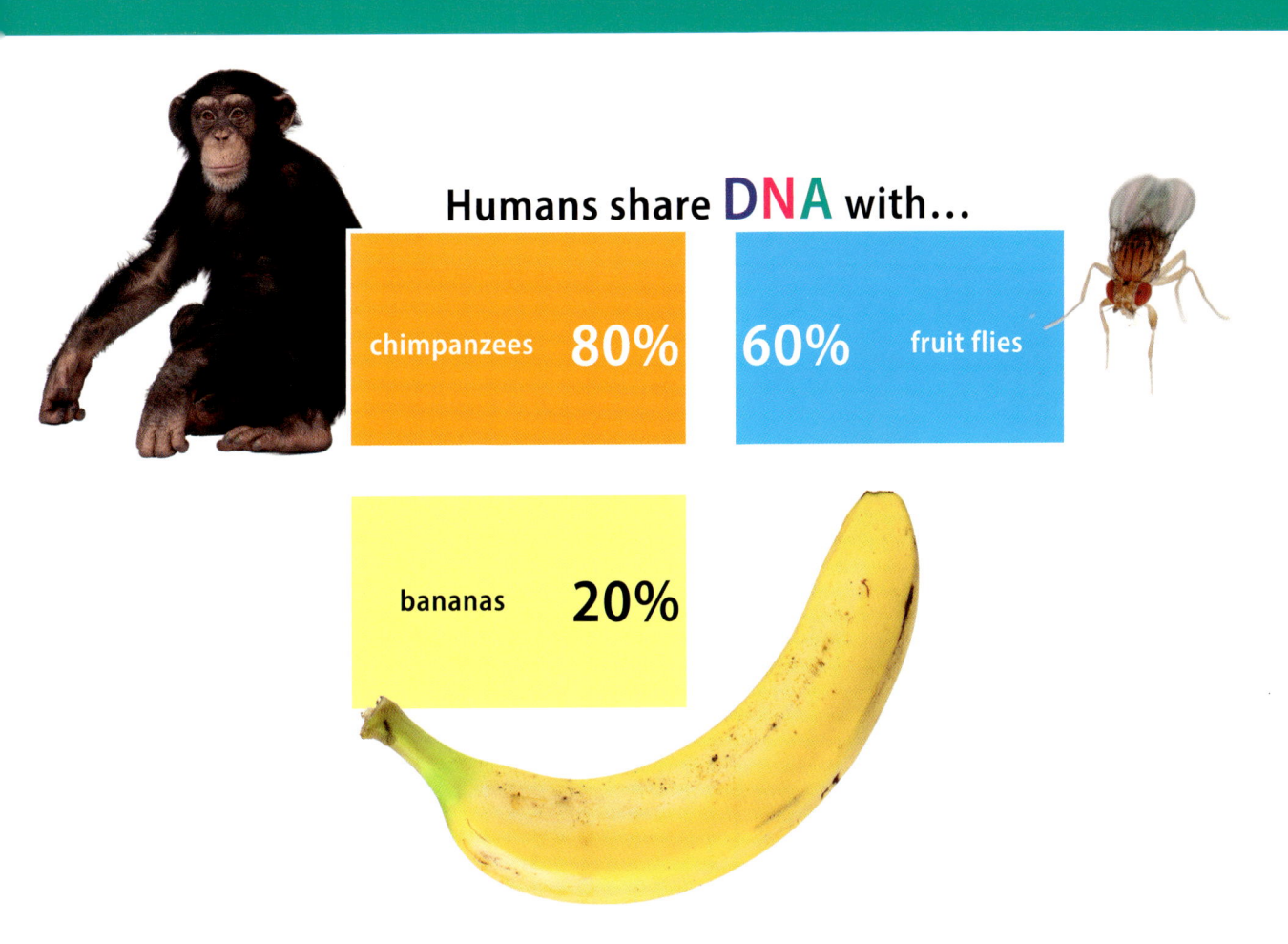

Humans share DNA with…

chimpanzees **80%**

60% fruit flies

bananas **20%**

Does that mean you are related to a BANANA?

No! God created living things with similar information so they could all live on the same planet. Think about it, if we need to breathe the same air and eat the same food, we should expect some of the same genetic information. The Creator God used a design code that worked over and over again in His creation.

https://www.genome.gov/about-genomics/fact-sheets/Comparative-Genomics-Fact-Sheet#:~:text=Comparative
https://lab.dessimoz.org/blog/2020/12/08/human-banana-orthologs

DNA in one cell contains a lot of information, equal to 10,000 novels of written notation.

If we could uncoil the 6 feet of DNA inside every cell of an adult and line it up end to end, it would cover the distance from the sun to Pluto and back again. That is how much organized information is inside every person. God is truly amazing!

Made in the image of God,
you are a special creation,
Created for a purpose,
you need Jesus for salvation.

Ephesians 2:10
For we are his workmanship, created in Christ Jesus for good works, which God prepared beforehand, that we should walk in them.

In other words...
We are made by God for a purpose. He created each of us to be unique and has prepared us for the purpose He has for us.

The DNA in your genes can tell you a lot of information: the places where previous generations of your family lived, who you are related to biologically, possible risks for certain medical conditions or diseases, and why you look like other people in your family. It's all stored in your DNA!

GRANDMA (Dad's Mom)

GRANDPA (Dad's Dad)

AUNT (Uncle's Wife)

UNCLE (Dad's Brother)

AUNT (Dad's Sister)

DAD

COUSIN

SISTER

BROTHER

46

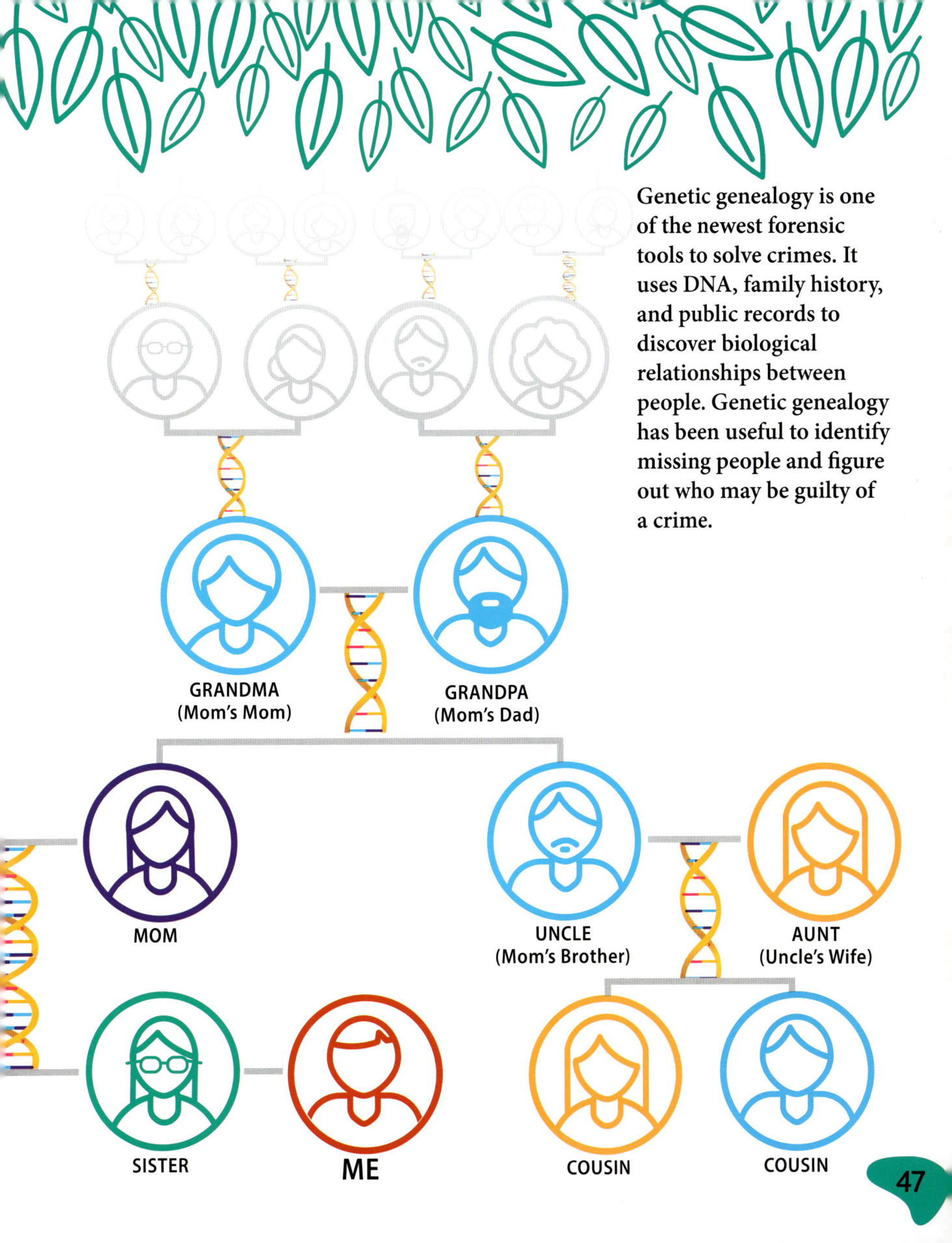

Genetic genealogy is one of the newest forensic tools to solve crimes. It uses DNA, family history, and public records to discover biological relationships between people. Genetic genealogy has been useful to identify missing people and figure out who may be guilty of a crime.

GRANDMA (Mom's Mom)

GRANDPA (Mom's Dad)

MOM

UNCLE (Mom's Brother)

AUNT (Uncle's Wife)

SISTER

ME

COUSIN

COUSIN

47

In the beginning,
God gave life to all things,
Unique DNA codes,
written by the King of kings.

I Timothy 6:13
God … gives life to all things.

In other words...
God creates and takes care of all living things, helping them to grow and flourish.

The Creator God spoke into existence the DNA code of life into all living things approximately 6,000 years ago. The orderly nature of DNA testifies to the orderly nature of God. You have been designed for a purpose, and God created your unique DNA code to reflect His glory.